IN THE FOREST

Photographs by
PETER DOMBROVSKIS
About Trees
JAMIE KIRKPATRICK

WEST WIND PRESS
Hobart Australia

PETER DOMBROVSKIS

Published and distributed by West Wind Press
PO Box 245, Sandy Bay, Tasmania 7006 Australia

Designed and wholly produced in Australia
by Rodney M. Poole Pty Ltd

ISBN 0 9586833 4 4

National Library of Australia Cataloguing-in-Publication Data:

Dombrovskis, Peter.
    In the Forest.
    Bibliography.
    ISBN 0 9586833 4 4.
    1. Trees.
    2. Nature photography.
    I. Kirkpatrick, J. B. (James Barrie).
    II. Title.
582.16

Acknowledgements

We would like to thank the following people for their
contribution to this book—Rod Poole, Alan Haig, David Ratkowsky,
Genevieve Gates, Safari Knight, Margaret Ricks and Di Williams.

CONTENTS

ACKNOWLEDGEMENTS   4

FOREWORD   6

ESSAY   9

POEM   32

THE PHOTOGRAPHS   —

TEMPERATE FOREST   33

TROPICAL FOREST   65

URBAN FOREST   81

BEYOND THE IMAGES

THE PAST AND THE FUTURE are two time continents, which we commonly separate by an ocean called the present. Does the present exist, or is it at best no more than a stopping point in time, gone before we can note it as present? We learn from the past and sail through the present in hopes of building a better future, a future in harmony with our natural environment.

As the Exhibits Chairman for the Photography Hall of Fame, I am constantly looking for those rare photographers who capture our senses and move our souls. Several years ago, while in London, I had the pleasure of viewing Peter Dombrovskis' work for the first time. I was mesmerised and moved to a long stillness to allow his images to seep into my soul. Peter portrayed the landscape with such radiance, exquisiteness, and composition that the pages seemed to speak to me.

I wanted to know more about this photographer and then I learned that he had recently passed away. During the following months, I sought to contact his wife, Liz Dombrovskis, and when I finally spoke with her, I immediately felt her love and passion for Peter, his art, and his message. Liz and I chose to share his message with others in the United States. With her help, and eighteen months of preparation, the Photography Hall of Fame Museum in Oklahoma exhibited Peter's work in the year 2000. At the opening, Australian Senator Bob Brown enlightened everyone with his stories of the images and of the adventures he had shared with Peter. The exhibit was one of the most admired and viewed displays in the museum.

In the year 2001, I made the journey to Tasmania and was invited to stay in the Dombrovskis home, where I met their children, and their friends. Peter's spirit seemed to surround the garden and the house, and emanate into the whole community. In the house were some of his images, with their simplicity and beauty, the depth of which enlivened many walls. Happily, I sat in his chair in the study and read his books, noticing his wide interests. He was highly educated in plants and landscape. He was passionate as a photographer, and was equally sensational as a gardener.

Every morning, I was allowed the pleasure of walking in the Dombrovskis garden—beautiful and serene, full of minerals and plants and gently flowing water. Brilliant vegetation rooted in the damp and fertile soil glistened in the sunlight, while shadows hinted at other mysteries and splendours. The haven was filled with a multiplicity of forms and was rich in individual creatures, all of which were to Peter of infinite value. None was preferred above the other; each in its own unique way had a place and function, value and beauty…even the weeds. Each was a signature of life.

Peter understood the relevance of the minute life in the natural environment. He captured in photographs, the individual beauty of many plants, in prints now on display in the Peter Dombrovskis Gallery at the Royal Tasmanian Botanical Gardens. This depth of beauty drew me to these images, and the extensive knowledge displayed in each caption again heightened my understanding of his passion for the wilderness. His lone journeys to seek out each magnificent species opened a door into a landscape of grandeur and an unfamiliar realm of seasons.

In every season there are beautiful moments—a rhythmic cycle of life where the whole forest moves seamlessly together. Animals, birds, and all types of creatures find homes. In many different ways, all forest was home for Peter, his solitude, and his sanctuary. His images speak to us about life and being alive. His images convey the forest's beauty and a message for all of us to keep: We should never exchange this beauty for money. Through his images, Peter gave the forest a voice when it did not have one.

Peter never considered trees to be static elements of nature. In some mysterious way, I believe he felt that we move through them and they seem to move through us. He loved walking among them, touching them, and smelling their fragrance. There is certainly something about their mysterious atmosphere, their silence, and their secrets that heightens the senses.

We love trees and we need them, we breathe what they exhale. They feed us when we are hungry. They comfort us when we lean on them. Their shade protects us and gives us shield. We build homes and raise our families underneath them. With them, we sail the oceans, discover and conquer new lands. They free our mind, heart, and soul from the loud and chaotic world. We use all faces of our friends…the trees.

The sacred journey begins with a seed. From this parched grain, which seems so utterly dead, a full life miraculously develops. The earth contributes its nutrients, the sky provides the water, even from the air the seed draws nourishment and from the sunrays it takes energy. Slowly the tree grows according to the definite laws of nature and the forest. Then comes the moment when the buds burst open and the flowers blossom; colors and shapes dancing in the wind, and in another moment the flowers start to fade. And now, in a mysterious way another seed from within will start the sacred journey and the eternal cycle will continue.

I feel so blessed to have merged into Peter Dombrovskis' universe. Although I never met Peter in person, I have entered his images, I have seen him being challenged at work, and I have met him there. I feel we are kindred spirits moved to photograph and speak for the natural world. A life without landscape denies the soul serenity, beauty, and diversity. Like Peter, for me a perfect day starts with the first inhalation of crisp, morning air, and the prospect of further photographic exploration of nature.

Peter was an independent, solitary photographer, who insisted on the quality of his photographic images so they could span a broader horizon of photographic vision. His love for and commitment to nature and its landscapes never ceased, they are now merely on a higher plateau.

YOUSEF KHANFAR

Yousef Khanfar is Exhibits Chairman, International Photography Hall of Fame,
Oklahoma City, United States of America. He is considered a highly respected photographer.
His work has been exhibited in museums and galleries worldwide, published in
magazines and in his book Voices of Light.

J. B. KIRKPATRICK

ABOUT TREES

I WOKE UP THIS MORNING to see a sky rendered red by dawn, through the blue leaves and flaked boughs of a spreading tree of blanket leaf (Bedfordia salicina). As I write, I contemplate a view through lawn, garden beds and trees to the coruscating waters of the Derwent Estuary. My remote ancestors sat around the campfire, washed in the sibilance of gently moving leaves and waters, eating shellfish and forest fruits; living in a similar scene. My not so remote ancestors engaged in the same activity in the same sort of place, recreating in their work/play dichotomised lives. I think that I am living now, in the now, in the act of writing in my ersatz cave in a contrived forest glen. Work or play—I do not know—perhaps neither—perhaps both.

Trees can be work. People plant them. People harvest their fruits and, occasionally, their leaves. People cut them down, either to create an ecosystem more suited to the satisfaction of their material desires, or to use their accumulated carbon for heat, structures, furniture, books, writing paper, disposable chopsticks, disposable packaging, newspapers and advertising supplements. The more people cut down, the happier is the very gross international product. It is very happy and getting happier.

At no time since humans started the work of cutting down trees have so many been felled or otherwise killed as in the past few years. In 1999 alone, more than 500,000 hectares of native vegetation was destroyed in Australia, most of it containing trees. In 1999 Australia was only the fifth ranking country, of those for which figures are available, for clearance in the world, and these figures do not count the cutting down of trees in native forest where natural regeneration was allowed to occur. There is still a lot of forest left, especially in the tropics and the taiga. In a world almost religiously committed to everlasting economic growth, our present record of tree destruction is likely to be transitory. Either so few trees will remain as to make such levels of destruction impossible, or the remaining forests will be respected and used, as forests, with care.

This book is not about trees as work. It is about the spirituality, beauty, biology and ecology of trees, the other reasons for keeping trees in the global landscape. The aesthetic and spiritual insights provided by the imagination of Peter Dombrovskis constitute the core of this volume. This text is about trees and their place in nature, gardens and the mind.

THE DIVERSITY OF TREES

TREES ARE ALL vascular plants. Vascular plants have developed a system for conducting water and nutrients through their tissues. Non-vascular plants have not developed this adaptation, so convenient in a terrestrial environment. Most trees belong to that group of vascular plants that produce flowers and form two cotyledons after germination, the dicotyledonous angiosperms. A few, such as palms (Plate 32) and pandanus (Plates 31, 34), are monocotyledonous angiosperms. The non-flowering group we know as conifers or gymnosperms contain a large complement of tree species. The ferns, or pteridophytes, have a small proportion of their species that are tall and woody enough to be considered trees.

There are many ways of placing the quarter of a million species of vascular plants that have evolved on our planet into groups. One way is by lifeform. The most widely used lifeform classification, that of Raunkiaur, uses the height of the perennating bud in relation to the ground as its primary criterion. The perennating bud is the location of new growth. For bulbs and tubers, termed geophytes by Raunkiaur, it is located beneath the ground. Trees fall into the phanerophytes, species that hold their perennating bud well above the ground. How high they need to be to earn the name tree is a matter of taste. Eight metres is a popular dividing line between shrub and tree. So is five metres. To get a perennating bud to these lofty altitudes requires some sort of reasonably solid superstructure. Thus, trees accumulate tissue in trunks and branches in the form of wood. It is generally considered bad form among trees to have more than one stem. However, this is tolerated, at least in Australia, when the perennating bud exceeds eight metres, while only the single-stemmed qualify between five and eight metres.

The most cursory glance upwards in a botanical garden will reveal many other characteristics that can be used to classify trees into lifeform. At the end of the growing season many trees release all leaves free. Autumnal colours are a precursor of this rejection. They denote the withdrawal from foliage of substances that will be recycled in the spring break, and a loss of connectivity with the sap of the tree. Evergreen trees stand bare only through misfortune or death. While leaves drop, often in concert, there are always live leaves left on the tree.

In temperate North America and Eurasia a large proportion of the forests and woodlands are dominated by deciduous trees, including many, such as maples (*Acer*) and elms (*Ulmus*), grown widely in the gardens of the Southern Hemisphere (Plates 47–49, 52). In the Southern Hemisphere most forests are evergreen, making colourful autumn scenes a rarity. In the temperate parts of the Southern Hemisphere there are some forests dominated by southern beeches (*Nothofagus*) that provide for a bright autumn. These are confined to southern South America and Tasmania. In the latter place deciduous beech (*Nothofagus gunnii*) gilds the autumn mountain landscape (Plates 11–13, 15–17). The southern beeches in New Zealand, mainland Australia, New Caledonia and New Guinea are all evergreen. New Zealand does, however, have winter deciduous species of lacebark (*Hoheria*) and fuchsia (*Fuchsia*). In both the Southern and Northern Hemisphere tropics, dry season deciduous trees can be found in both forests and woodlands.

Deciduousness has some apparent disadvantages in the internal economy of trees. The production of leaves requires inputs of considerable energy and material resources. Their rejection after but one growing season would seem less efficient than retaining leaves for several years (the world record for a tree is forty-five years for needles of bristlecone pine (*Pinus longaeva*)). While deciduous trees are leafless there is opportunity for evergreen trees to capture light denied them during the leafy season and thereby gain a height advantage that would deny their deciduous companions light. The widespread dominance of

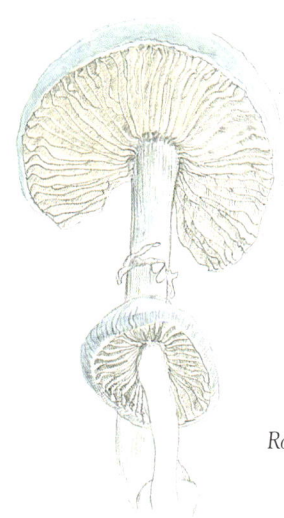

Rozites metallica

deciduous trees suggests that these disadvantages are outweighed by advantages conferred by total leaf loss during the winter or dry season.

Leaves produced for just one growing season may not involve as great an investment per unit of photosynthetic return as more persistent leaves. Old leaves accumulate grime and are sometimes partly covered by non-vascular plants, becoming less photosynthetically efficient than young leaves. Also, leaf demise is not necessarily voluntary. Leaves meet premature death as a result of predation by animals and attacks by fungi and microorganisms. Evergreen trees have been under severe selective pressure to reduce this leaf loss sufficiently to allow their growth and reproduction in competition with other trees. Consequently, they typically invest in protection, such as through the production of poisons, spines and tough lamina. In a deciduous forest the predators and parasites of leaves are doomed to a very lean season. These consumers of foliage have evolved to recover their numbers once their resource becomes again available, but the trees tend to have time to grow their leaves before the situation becomes dire, soon after which time they can drop them again. This strategy would not be of much use in environments where non-deciduous plants were sufficiently abundant to carry predators and parasites over the leafless seasons. This may be part of the reason why deciduousness is generally associated with environments that have a season that is highly unfavourable to plant growth. The occasional evergreen plant in a deciduous vine thicket in the tropics puts on little fresh growth in the dry season, as there is a marked lack of available moisture. In areas where the soil freezes in winter or is covered by deep snow, the evergreens that occur cannot make much cold season growth. Those that do grow a little tend to have invested substantially in protection, which they need when they are the only eating option in a time of starvation. Their consequent slow growth tends to prevent dominance over the deciduous trees.

There is seldom one solution to the same problem in nature. Evergreen trees do occur in the same climatic environments as deciduous trees, where they are often dominants or can occur in intimate mixture (Plate 12). A common pattern in the northern temperate climatic zone is for deciduous trees to be dominant on the more fertile soils and the evergreen trees, usually pines or other gymnosperms, to be dominant on poorer soils. A massive addition of leaf area in spring may require a more rapid movement of nutrients from litter to feeder root than can occur on acid, nutrient-poor soils.

In the particular case of the Tasmanian deciduous beech, these theories of deciduousness seem to have somewhat limited application. The richly aromatic nature of its leaves suggests that it may have had to evolve protection, despite its deciduous habit. Deciduous beech is usually well mixed with evergreen trees, shrubs, grasses and herbs, and never far from them, a situation consistent with this hypothesis. Deciduous beech is common on some of the poorest soils in the world, shallow peat mixed with quartz gravel over quartzite. It is more common where snow lies relatively long and winter sun seldom penetrates, than elsewhere, but the maritime Tasmanian climate makes mountain cold and snow a short term phenomenon in winter.

Photosynthetic organ size, shape and arrangement are other attributes widely used in lifeform classifications. A small part of the range of variation can be seen in the needles of Monterey pine (*Pinus radiata*, Plate 57), the tiny leaves of myrtle beech (*Nothofagus cunninghamii*) arranged alternately along the stem (Plate 23), the long hard serrated leaves of pandani (*Richea pandanifolia*) forming a terminal rosette on a long bare stem (Plates 2, 4, 5), and the large leaves of tropical rainforest trees (Frontispiece) and *Rhododendron* (Plate 55).

There is some general tendency for leaves to become smaller as climate becomes colder. Small leaves may be more efficient than large in minimising water loss per unit of photosynthetic production. However, this theory does not explain the tendency to smaller leaves in colder climes. With both drought and frost damage, the parts of leaves most remote from their internal water distribution systems die first. In a small leaf, such remoteness may be minimised.

Small trees that are largely confined to the understorey of taller trees tend also to have larger leaves than are found in the forest canopy. The penetration of light energy to the understorey of a rainforest is typically only one percent of that incident on the forest canopy. In such shady conditions, where light arrives as flecks and diffuse radiation, a large leaf may provide enough energy to make it work efficiently as a photosynthesis factory, whereas a small leaf might fail.

Not all trees use leaves for photosynthesis. Some, such as blackwood (*Acacia melanoxylon*), lose their leaves as seedlings and the leaf stalk expands to fulfil this function in an organ known as a phyllode. Others, such as the sheoaks (*Allocasuarina*), photosynthesise in branchlets, known as cladodes, with the leaves left as vestigial brown scales.

TREES AS PRODUCERS

LEAVES, PHYLLODES AND CLADODES exist to photosynthesise. To undertake this productive task, on which many more organisms than trees depend, they need water, carbon dioxide and light.

Trees with foliage in the forest canopy do not generally suffer from a deficiency of light energy. Their major problem is akin to the problem of balancing heat loss and access in buildings, solved by the revolving door. To let in carbon dioxide trees must release water. Carbon dioxide is everywhere in the atmosphere, in greater quantities now than several centuries ago. Water must be obtained from the soil, and can easily be exhausted. Desiccation means death. The bark of the tree and the skin, or cuticle, of the leaf let little water escape. To allow the access of carbon dioxide there are small holes in leaves, called stomata, opened and closed by guard cells. Most of the loss of water from trees, known as transpiration, takes place when these stomata are open.

THE FORM OF A TREE

THE FORM OF A TREE is partly inherent and partly environmentally induced. The umbrageous myrtle beech in Plate 18 may have had the more erect form of the myrtle beeches in Plates 7 and 8 if other myrtles had managed to establish nearby. The tree form evolved to capture light in competition with other plants. Trees preferentially grow towards the light. If light is available in all directions they spread (Plates 3, 13, 18, 38, 41, 42, 48). If light shifts in its location as they grow, their trunks sinuate (Plate 30). If a dense stand of seedlings establishes after adult trees are killed by a disturbance, such as fire, lateral light will be limited and the trees will grow relatively straight and tall, as for the satinay in the Fraser Island forest (Plate 30) and the swamp gum (*Eucalyptus regnans*) in the Dombrovskis garden at Fern Tree (Plates 55, 56). However, tree species and individuals vary genetically in their propensity to grow straight and tall. When grown together in the same conditions swamp gum will be a straighter and taller tree than cabbage gum (*Eucalyptus pauciflora*, Plates 28, 29).

GETTING WATER TO THE CROWN

THE VERTICAL DISTANCE between the root system and foliage of many trees exceeds the monumental, with living trees of Californian redwood (*Sequoia sempervirens*), and demised trees of swamp gum, well exceeding one hundred metres from ground to their highest foliage. The theory that presently best explains how water and nutrients make such great vertical movement involves a mechanism that commences in the intercellular spaces of the leaf during the process of transpiration. The water that moves through the membranes of the cells to be lost to the atmosphere makes the dissolved substances in the cell more concentrated. Water then moves into the cells from below, pulled by the difference in solute concentration. Because water is not very easily pulled apart, requiring in the order of seventy times the force required to get it to the top of the tallest of trees, it moves from the roots up into the foliage through vessels in the trunk. Water is then pulled from the soil into the roots to keep the whole process going. The movement of water up the stem is aided by the ability of the walls of the spaces through which it moves to hold water tight. This water adherent quality and the narrowness of the spaces through which water moves militate against the possibility of any potentially fatal breaks in the water column (21).

HOLDING IT ALL UP

THE ABILITY OF TREES to occupy space well above the ground not only depends on a mechanism to ensure transport of water and nutrients in a vertical direction, but also on the possession of the superstructure provided by roots, stems and branches. If trees were not able to expand this superstructure laterally as well as vertically they would become unstable at diminutive statures. Lateral expansion has been termed secondary growth. Variation in the rate of secondary growth is responsible for the ringed structure in wood that can often be used to calculate the ages of trees.

THE AGE OF TREES

THE INDIVIDUAL LIVING tree stem that is presently accepted as the oldest on the planet, on the basis of counts of its growth rings from a core obtained from its base, is a bristlecone pine that has grown near the treeline in the White Mountains of California for at least 4900 years. However, there is a strong possibility that older genetic individuals of trees may exist, with their stems shifted in space. On Mount Read in western Tasmania there is a small stand of Huon pine (*Lagarostrobus franklinii*), a species whose stems are known from ring counts to live for more than 2000 years. We know from analysis of the pollen deposited in nearby Lake Johnson that the species has been at this site for at least the last ten thousand years. We know that Huon pine can produce new stems from root sprouts. We know that the population of Huon pine on Mount Read consists entirely of males (23). We do not know the duration of this rather marked sex bias, which mirrors that of humans in some of the nearby mining towns, but can speculate from the above set of information that a single individual of Huon pine may have survived on Mount Read for ten thousand years. Another Tasmanian tree species that is known to have single stems that have lived for more than a thousand years, and which produces new stems from root sprouts, is the pencil pine (*Athrotaxis cupressoides*), Plates 19, 21. At subalpine altitudes it is common to see almost dead huge trees of pencil pine with lines of progressively smaller trees radiating from them (6). Fires that have killed the copses and incinerated the organic component of the soil have revealed the subterranean connectivity of the apparent individuals.

Boletus sp.

TREES AS KEYSTONE SPECIES

A KEYSTONE SPECIES is one whose removal would have negative consequences for many other species. Tree species are often keystones, and are almost always a keystone lifeform, given that they control so many aspects of the environment in which shorter species must exist.

SHADING

IN GROWING TALL to capture light, trees create shade. The nature of the shade strongly influences the nature of the understorey. In deciduous forests there is a profusion of species that take advantage of the relatively brief periods in early spring and late autumn when canopy cover is not dense, but temperatures allow growth. Bulbs, such as bluebells (*Hyacinthus*, Plate 45), and perennials, such as anemones (*Anemone*) and lenten roses (*Helleborus*) colour the forest floor. Few plant species can attain much cover in the dense shade of the evergreen rainforest canopy (Plates 3, 7, 8, 10, 33, 37–39). In contrast, most species of eucalypt let a generous amount of light through to the next stratum in the forest. Consequently, they often develop a species-rich understorey of substantial cover (Plates 28, 29). Shady places are cooler than those fully exposed to the sun. Mean daily maximum temperatures beneath a relatively open canopy of swamp gum (Plates 55, 56) have been shown to be 2°C cooler than in adjacent open areas (26). Shaded places are also relatively warm when open places are crusted with frost or dusted with snow (Plate 22), the canopy of the forest reflecting downwards the heat that would otherwise escape from the soil to outer space. For example, in clear sky conditions minimum daily temperatures can be 6°C higher under an intact eucalypt canopy than in the open (26). This moderation of temperature conditions allows plant species to occur beneath trees in places where they could not survive in the absence of trees.

EFFECTS ON MOISTURE

THE EFFECTS OF trees on the moisture available to subcanopy plants is more complex than their effects on temperatures. Trees can capture moisture that would otherwise not precipitate. In subalpine eucalypt forests, on mountainsides subject to frequent mist and fog, there is a capture of three to seven percent more moisture than is the case in the absence of trees (26). Fine water droplets, which would otherwise drift past to ultimately evaporate, coalesce on leaves, branches and trunks, then drip or flow down trunks to the soil. Snow can also be captured by isolated trees or copses of tree, both immediately (Plate 19) and in the lee of the wind breaks that they form. However, a substantial proportion of precipitation is intercepted by tree canopies, to be evaporated before it can reach the ground. The proportion of rain thus intercepted has been measured to vary from one-tenth to one-quarter in eucalypt forests (26). A few percent of rainfall at canopy level ends up reaching the ground as stem flow, making the soil at the base of trees slightly moister than other soil directly under the canopy (26).

Trees, constituting as they do the major proportion of the biomass of a forest, use large amounts of soil moisture, rendering the soil relatively dry for other species wherever the surface mat of tree roots spreads. In relatively dry eucalypt woodlands, where trees are more widely spaced than in forests, the extent of the root mat can be up to thrice the diameter of the tree crown. In wetter eucalypt forests and rainforests the root mat is continuous in the absence of tree death or fall, but moisture is not as often a limiting factor for the establishment of understorey species.

Transpiration by trees can also be important for other species, in that it can prevent waterlogging or salinisation. In the nineteenth century in California there was a widely accepted theory that malaria was caused by miasmas drifting from swamps. The volatile oils released from eucalypts with transpiring moisture were thought to be effective in counteracting these miasmas. Consequently, swamps were surrounded with plantations of Tasmanian blue gum (*Eucalyptus globulus*). Despite being based on an incorrect theory of causation of the disease, this prophylaxis proved moderately effective. The transpiration of the rapidly growing blue gums lowered the water table, eliminating the swamps and the mosquitoes that lived within them. The removal of trees results in an increase in the amount of moisture reaching the water table. If the water table reaches the soil surface where there is no salt in the subsoil, plants resistant to waterlogging will replace those adapted to well-drained soils. Where there is salt in the subsoil, a white desert is the consequence of tree loss.

The replacement of a mature forest with a rapidly growing young forest also has consequences for other living things. Young forests use much more moisture than old ones. This affects the amount of water available for stream flow (2), with consequent impacts on stream biota.

EFFECTS ON WIND

THE FLOOR OF a forest is generally a more calm and humid place than the canopy or adjacent treeless places. Trees slow down the movement of wind by providing frictional resistance. Thus, water transpired in the understorey or evaporated from the soil tends to accumulate as vapour in the subcanopy, rather than being whisked away. Rainforests with their dense canopies tend to have especially calm and humid air, providing the habitat for such hygrophilous (humidity-loving) plants as the diaphonous filmy ferns, which commonly nest among the mosses on trunks and fallen logs in the types of temperate rainforest depicted in Plates 1–7. Eucalypts abate the wind less. In Tasmanian snow gum (*Eucalyptus coccifera*) forest on Mount Wellington, the strongest winds have been shown to be unaffected by the presence of trees, which still calm, to some degree, less ferocious air movement (11).

EFFECTS ON NUTRIENTS

TREES ARE nutrient pumps as well as water pumps. They generally have the deepest root systems of any plants in a forest or woodland, thereby being able to access stores of nutrients not available to other plants. These nutrients eventually find their way to the forest floor, partly in the form of tree litter, which is broken down by other organisms, then recycled by trees and other plants (2). The litter of different tree species varies markedly in both the generosity of its nutrient content and its balance between various types of nutrients. For example, the leaf litter of deciduous trees, such as elms (Plate 52), is both richer in nutrients and less acid than the leaf litter of pines (Plate 57). When pines are planted on the brown soils typical of deciduous forests they eventually turn the surface soil grey, the leachates from their acid foliage and litter carrying iron and other nutrients down the soil profile. Plantings of deciduous trees can reverse this process.

As well as the major contribution of leaf litter, nutrients also move from tree to soil dissolved in water that has flowed through foliage and down trunks. Downward movement can also be in the form of twigs, bark, branches and whole trees, or as the waste products of animals that live off, on or in the trees. These include corpses, frass (insect wastes), scats (vertebrate droppings) and urine.

LITTER

AS NOTED ABOVE, the nature of litter on the forest floor differs markedly between different types of forest, and this has major ecological consequences. In tropical rainforest, heat and moistness conspire with living decomposers, such as fungi and invertebrates, to break down litter at an extremely rapid rate, leaving scattered disintegrating leaves that barely conceal the soil (Plates 33, 38, 39). The lack of understorey plants in this context cannot be attributed to smothering by litter, rather being caused by dense shade and root competition. There is insufficient fuel for fire to spread. This is a considerable virtue given that rainforest is easily destroyed by fire (1, 5). There is also very little protection against rain splash erosion, which perhaps accounts for the razor-like ridges often found supporting tropical rainforest.

The floor of the temperate rainforest often has a twig-rich litter (Plate 7). Breakdown is slower than in tropical rainforest, because of cooler temperatures. This litter protects the soil from erosion, but will support a slow, creeping fire, which can kill the trees by basal ringbarking. Fallen trees and large branches are more common and persistent in these forests than in tropical rainforest (Plates 1, 6, 7, 8, 10 compared with Plates 33, 37–39). These play an important role in the regeneration of tree species (19), the seedlings of which are usually absent or sparse on the forest floor. Seedlings are unable to establish in competition with the dense root mats of adult trees, under relatively dense litter and with the constant disturbance of animals seeking food, but can establish on the fallen timber. It is common to find trees in straight lines in such forest, indicating the location of a long disintegrated fallen tree. It is also common to find trees whose arched roots join the trunk above ground level, subtrunk emptiness denoting the past location of the log or stump on which the tree established.

The litter in eucalypt forests (Plates 27–29) is typically much more profuse and varied than that in any rainforest. Eucalypt leaves tend to resist breakdown to a greater degree than those of most rainforest species. They are rich in aromatic and volatile oils, are acid and have a large component of hard tissue. Eucalypts also produce more leaf litter than most rainforest trees in similar climes, having superior growth rates. Eucalypts not only profusely shed leaves; many species also prodigiously shed bark, much of which finds its way to the base of the tree (Plates 28, 29). Neither are they reluctant to drop branches. Consequently, large amounts of litter accumulate at the bases of eucalypts, which, if not burned away in the frequent fires that take advantage of the fuel, slowly break down into a fibrous woody substance known, somewhat unromantically, as duff. The duff mound at the base of old swamp gums in long unburned forests can be metres high. Like the deciduous bark, strings of which can be carried alight for kilometres during a fire, the flammable duff pile increases the chances of the burning required for most eucalypt regeneration. This helps prevent the inevitable victory, in the prolonged absence of fire, of shade-tolerant rainforest species over the sun-loving eucalypts.

Aristotelia peduncularis

HABITAT FOR BIRDS

TREES FORM A large part of the base of the ecological pyramid; they are among the producers that support both consumers, like us, and, eventually, decomposers, like fungi. They also interact with other organisms in a myriad of ways outside the satisfaction of appetite.

Birds and trees have a particularly intimate reciprocal relationship. Some biologists recently consumed some scarce research time to test the hypothesis that it was safer to be in the air than on the ground. Despite attacks by raptors and the possibility of plane crashes, their observations were consistent with the hypothesis. Being fully in the air is fine for travel, hawking and, occasionally, sex, but leaves something to be desired for egg incubation, hatching and juvenile care. Trees are the compromise solution for a large number of bird species. Nests can be constructed in the forks of branches, dangling from the same, or within hollow branches or limbs. Old, slightly unhealthy trees provide some of the best nesting opportunities, hollows and height not being defining characteristics of youth among either humans or trees.

Being a bird condominium has some advantages for trees. Most of the food consumed by the residents, and disposed of by gravity to enrich the soil, comes from elsewhere than the tree itself. Birds are generally not great leaf eaters. They may strip a few for enjoyment, as the white cockatoos appeared to do in the Norfolk Island pines I can see from my study window, or may industriously use a few in nest construction, but their major foraging activities involve invertebrates that suck sap, or that consume leaves in prodigious portions. Hugh Ford estimated that 55–70 percent of invertebrates may suffer premature deaths in the beaks of birds in eucalypt forests (26). The consumption of such quantities of invertebrates can only aid tree health and well-being. Of course, if the residents are bird-eating raptors, such as wedge-tailed eagles, the invertebrates can munch and suck unpredated, a reason why trees often die in close proximity to raptor nests.

FOOD FOR ANIMALS

MANY OTHER VERTEBRATES that frequent tree tops are much more prone to consume foliage than birds. Only the Fijian forests depicted in this volume would be free of native leaf-eating arboreal mammals, and even those have tree-climbing introduced rats. Large numbers of both birds and mammals appreciate the nectar and fruit produced by many trees. If trees could appreciate, they might appreciate the appreciation in the chances of their genes being perpetuated through the pollination and dispersal services they gain from these appreciative birds and mammals. They have certainly evolved to encourage such services. The nutrient-rich large seeds that compose most of the football-sized infructescences of *Pandanus* (Plate 31) have been interpreted to be an adaptation that encouraged dispersal by the now extinct Pleistocene megafauna. Such interpretations can only be speculative. For example, megafaunal maws might be thought to be necessary to gain access to the white flesh and sweet milk of the coconut (*Cocos nucifera*, Plate 32). However, the effective dispersal of this species has been by sea currents, which have carried it successfully to coral atolls that have never felt the pressure of megafaunal feet.

Trees offer many lifestyle opportunities for vertebrates, apart from habitation. Folivores, such as the ringtail possum and the koala, do the hard work of digesting fibre-rich leaves full of chemical defences. Seed eaters, such as rosellas, tear apart cones and woody fruits to gain sustenance from seeds. The only benefit for trees, such as eucalypts and pencil pine, that lose some of their reproductive potential to this typically raucous dining activity, is some accidental dispersal of seed away from the parent tree.

Frugivores, such as pigeons and flying foxes in tropical rainforest, and currawongs in temperate rainforest and eucalypt forest, do trees more of a favour in return for the food the trees provide. Many tropical rainforest trees, and some temperate rainforest trees, have evolved fleshy coatings or appendages that become palatable to vertebrate dispersers when the seed is mature, and have seeds that either survive, or have their dormancy removed, by passage through vertebrate digestive systems. The most common gymnosperm in Tasmanian rainforest is the celerytop pine (*Phyllocladus aspleniifolius*). It has a delicious succulent attachment, or aril, firmly adhered to an indigestible seed. It is consequently well-dispersed by currawongs, often being found growing beneath places on which they perch (3).

Nectivores also do the tree a favour, dispersing pollen within trees and from one individual to another, thereby facilitating outcrossing. The nectar is the reward for this sexual activity, which is engaged in vicariously by birds, bats and non-flying mammals. All this somewhat devious investment in sexual activity has been theorised to have evolved in response to disease, which could potentially wipe out all individuals of a species if there were not much variation in individual genotypes.

Insectivores have adopted a variety of feeding styles that take advantage of the peculiar qualities of trees. Avian hawkers and flitters hawk and flit between trees at dusk, in daylight and at dawn, capturing flying invertebrates from arboreal launching pads. Bats take the night shift. Some birds, such as yellow robins, pounce rather than hawk or flit.

Other birds are less airborne in their feeding activities, gleaning invertebrates from leaves, branches or trunks, or engaging in deadly probing beneath bark. While insectivores generally benefit the trees they work, as well as themselves, those birds who declare trees private property could be a cause of tree death.

Miners are aggressive medium-sized insectivores who are extremely jealous of any predation of invertebrates by other birds. Where they establish colonies they chase away anything chasable. Miners' eyes must be bigger than their stomachs, as tree dieback, proximally caused by invertebrate damage to foliage, is often associated with their colonies.

The number of species of vertebrate that depend on trees is large. The number of invertebrate species that depend on trees is enormous. In a rather scrubby rainforest in Panama, Erwin obtained almost 1000 species of beetle alone, from just nineteen trees of the same species sampled over three years (25). This compares somewhat favourably with the 750-odd species of vertebrate found in probably the most vertebrate-rich terrestrial part of the planet, the rainforest of Cosha Cashu in Peru (25). Harry Recher recorded 727 invertebrate species from the canopy of just one eucalypt species, narrow-leaved ironbark (*Eucalyptus crebra*) (26). An average figure of ten individual invertebrates per square metre of foliage has been calculated for trees other than eucalypts in southeastern Australia, with eucalypts supporting, on

Pseudaeolus australis

average, eight invertebrate individuals per square metre (26). While not all invertebrates are herbivores, the vegetarians among them eat a lot of tree.

The average mature leaf on a tree in a forest or woodland has paid a tithe of approximately ten percent of its area to the invertebrate world (for instance, Plate 27). This tithe is largely exacted by leaf-chewers. In eucalypt forests these are typically beetles, stick insects and sawflies; the larvae of the latter of which, during the day, collect in black balls that writhe and exude yellow poison when disturbed. The leaf miners, largely caterpillars, excavate holes in leaves. Skeletonisers are more particular in their tastes, removing only the softer parts of the leaf. The tithe is not constant. Christmas beetles and chrysomelid beetles are renowned defoliators, often leaving behind nothing, while some branches can remain almost entirely insect-free and unscathed (Plate 23). The tithe is not even all that is lost, because it does not account for totally consumed leaves and leaves that drop as a result of extreme damage.

Sap-suckers are not usually part of the leaf area tithe, but discolour or kill the parts of plants on which they feed. Nevertheless, in extreme cases they can cause severe defoliation of trees. The most notable of the sap-suckers in Australia are psyllids. In their nymph stage they conceal themselves in a starchy shell, known as a lerp. These lerps attract gleaning birds and ants, making them perhaps not the best hiding place.

Not even the superstructure of trees is sacrosanct to invertebrates, many of which are wood-borers. However, most wood-boring activity takes place in dead wood or damaged live wood. Termites, long-horned beetles, ghost moths and wood moths are the main groups that bore eucalypts. Dead wood on myrtle beech is bored by the *Platypus* beetle, once blamed for myrtle dieback but now known to be innocent, the guilty party being a fungus, *Chalara*.

Some insects, including wasps, scale insects, flies and psyllids induce abnormal growth on trees, which provides sustenance for their young. These galls can vary from inadequately enclosing leaf rolls, to imaginative balls decorated with spines. They can be found on all parts of the tree, those in the reproductive area sometimes inducing inexperienced observers to think that they have discovered new genera of plants.

There are also invertebrates that eat flowers, fruit, nectar, seed, bark and fine roots, in fact, any and all parts of a tree. While there is no doubt that herbivorous invertebrates are occasionally arboricidal, they do provide some benefits in some circumstances. The bizarre product of coevolution between fig wasps and figs (*Ficus*) provides an extreme example, active in the types of forest shown in Plates 33, 36–41. There is a different fig wasp species for each species of fig. They develop within the ovaries of gall flowers that are within some fruits of the fig. Once sexually mature, they chew their way out to a more spacious part of the interior of the fig. Males and females mate within these spaces. The male wasp, which lacks functional wings, remains within the only fruit it has known. The female leaves the scene of her impregnation through a convenient aperture known as the mouth of the fig, brushing against male flowers in the process. These cover her with pollen. She then flies to another tree, lured by the scent of young fig. If this is the type of fig that has female flowers, these are pollinated as she searches for gall flowers, seed is eventually formed and the procreation of the fig is well underway. The literature is not clear on what subsequently happens to the female wasp. I assume that she emerges in search of a fig with gall flowers. Once attaining the interior of a fig with the right type of flower she inserts her ovipositor and lays an egg in the ovary—and so the generations of fig wasps and figs roll on.

Pollination is the big plus contributed by invertebrates who eat trees. However, other activities are not necessarily totally negative. Inasmuch as hollows in trees contribute to tree well-being by attracting vertebrates, the role of wood-eating termites is positive. A heavy to severe pruning can also make some trees more healthy than they otherwise may have been, especially in gardens.

On a dewy morning, looking into the sun through Australian trees, it is best not to suffer from arachnaphobia, as webs sparkle between almost every branch. While not generally as dependent on trees as the many invertebrates that can eat nothing else, invertebrates who eat invertebrates often depend for their prey on the presence of trees. Trees can also depend on these predatory animals to protect them from severe outbreaks of herbivores. The classic example of this is the relationship between the cottony cushion scale and the ladybird in the citrus orchards of California. The cottony cushion scale, an accidental Australian import, threatened citrus production in California, until suppressed in numbers by the predatory ladybird, also from Australia (18).

TREES AS HABITAT FOR OTHER PRODUCERS

TREES NOT ONLY provide resources for species higher up the food chain. They also provide habitat for other producers. Some of the most spectacular of these are climbers. One type of climber winds its way up to the canopy where it garnishes to overwhelm the tree tops. These climbers can develop spectacularly large stems, in excess of the diameter, and in deficit of the pliability, useful for Tarzan (Plate 33). Strangler figs (Plate 36) start out life as such climbers, ending up as hollow centred trees, the hollows marking the space once occupied by the trees they enveloped. These big woody climbers are highly light demanding, so tend to be most common in forests damaged by wind throw or on the forest margin. The other type of climber attaches itself firmly to boles using specialised roots (Plates 39–41). Many of these bole climbers are relatively tolerant of shade. Some die off behind as they climb, converting themselves into epiphytes, plants that grow on, but not off, other plants.

In 1978 Dave Hassall and I recorded all the vascular epiphyte species on more than 300 trees on Mount Korababa near Suva in Fiji (12). In the deep valleys, where the rainforest was tall with many palms and lianes, there were more than ten species of epiphytes on some single trees. On the exposed razor-back ridge that constituted the peak of the mountain, the relatively short and small-leaved rainforest trees supported far fewer species of epiphytes. The types of epiphytes varied with both environment and position on tree. The epiphytes that grew in the crowns of the trees tended to have thick leaves which resisted desiccation, whereas those growing in the constantly humid subcanopy were thinner leaved. Nest-forming ferns (Plate 33) were common in the valley forest. These trap detritus, creating their own organic soil which they then explore with their roots. On the ridge an orchid which consisted of a small circle of a leaf pasted to the bole of the tree was relatively common. Other more statuesque orchids, such as those in the genus *Dendrobium*, had swollen leaf bases, known as pseudobulbs, in which they stored water between rains.

One of the more interactive epiphytes in the Mount Korobaba rainforest was an ant plant (*Myrmecodia*). These plants develop swollen stems riddled with internal chambers, well suited to occupance by ant colonies. The presence of an ant colony may fertilise the plant and protect it against invertebrate predation (26).

Epiphytic plants live in an extremely droughty environment, with no or minimal soil to store water between rains. This is why there are epiphytic cacti in some Central American rainforests, why epiphytic species increase from two percent of the flora in relatively dry tropical forests to more than a quarter of the flora in the wettest of tropical forests, and why epiphytes make good house plants (26).

Trees, climbers, orchids, ant plants and ferns are all vascular plants. Non-vascular plants also colonise trees. There are fewer species of non-vascular plant than vascular plant that occur in forest. There are approximately 250,000 species of vascular plant, almost all of which are terrestrial and a large proportion of which occur in forests. The mosses, liverworts and hornworts, collectively known as bryophytes, constitute only 16,000 species. The other main group of non-vascular producers that is found in forests is the lichens, of which there are approximately 20,000 described species (21).

Lichens are regarded as honorary plants, because they photosynthesise. However, they are a parasitism to weak symbiosis deemed to be an organism, consisting of a fungal superstructure and a photosynthetic component provided by green algae or cyanobacteria. Their fungal component is more closely related to animals than to those other organisms we label plants. For example, their cell walls are not constructed of cellulose, but rather of chitin, the material that composes the shells of crustaceans. The fungal component of lichen will survive in the absence of the algal or cyanobacterial component, but adopts a very different shape and loses the carbon and nitrogen it gains from them. The algae and cyanobacteria have been shown to be more productive in the absence of the fungal component, and produce different outputs when within the fungus than outside it. Lichens adopt a wide variety of forms. The crustose form can be seen on the plate buttress roots of the tree in Plate 37. A fruticose, or leafy, lichen can be seen in Plate 6. A hanging lichen, probably a species of *Usnea*, can be seen in Plate 35. A suffruticose, or shrubby, lichen grows on a banksia cone in Plate 27.

Whereas moist tropical forests are notable for their abundance of vascular climbers and epiphytes, the temperate rainforests of Tasmania have relatively few. These are largely epiphytic filmy ferns (*Hymenophyllum*) supplemented by a few climbers. The pink-flowered climbing heath (*Prionotes cerinthoides*, Plate 4), a Tasmanian endemic, is the most frequently encountered climber in the montane rainforests of western Tasmania.

While bryophytes and lichen are found as epiphytes in tropical rainforest (Plates 33, 36–41), they are most profuse in the temperate rainforest (Plates 1–8, 22), where 55 species of bryophyte and 76 species of lichen have been recorded by Jean Jarman and Gintaras Kantvilas from a single tree of Huon pine (14). The various tree species in the Tasmanian temperate rainforest differ in their propensity to develop a dense covering of epiphytes on their trunks and branches. For example, there is a much greater non-vascular cover on the trunks of myrtle beech than on those of King Billy pine (Plate 1) or deciduous beech (Plate 11), possibly relating to variation between species in deciduousness of bark, roughness of bark and chemical composition of bark. There is also substantial variation in both cover and species composition within individual trees (Plates 1–3, 6, 7), with both total cover and bryophyte cover being highest in moist shaded situations and lichen cover being more prominent on the drier parts of the trunks and branches.

Epiphytes and climbers may smother trees with their foliage, crush them with their stems or break their branches with their weight, but have the grace, albeit only by definition, not to feed off their living tissue. Parasitic plants have evolved to be able to extract moisture and nutrients from living trees using specialised roots called haustoria. The mistletoes are tree-dependent parasites, common on the mainland of Australia, but absent from Tasmania. Their sticky seeds are dispersed, not surprisingly, by mistletoe birds, in the act of cleaning their beaks after snacks.

Prionotes cerinthoides

DECOMPOSERS

THE HAUSTORIA of some species of fungi consume living tissues. However, more fungi are into decomposition than parasitism. Decomposers do not have quite the same glamour as either producers or consumers, but without their rotting the world would be a very different place. The biosphere is finite and does not change much in the magnitude of its supply of the constituents of life. The faster the elements of life can cycle, the greater the growth of producers and consumers that can be supported. Fungi, ably supported by some invertebrates and many microorganisms, take dead bits of trees and break them down to their elemental particles, which then can be reconstituted into live tree. Fungi do not incorporate their dead products in their living self, there being no equivalent of toenail or heartwood. Most fungi, including mushrooms, consist of filaments of cells called hyphae, a mass of which is called a mycelium. Hyphae can be hyperactive in their growth, a kilometre of extension in a day not being regarded as outrageous for a single individual fungus. The fungus has little of its material remote from the rest of the world. It produces enzymes that dissolve this world around it. There are approximately 100,000 species of fungi that have been scientifically described, and probably as many again awaiting taxonomic attention. Therefore, the fungal world is almost as rich in species as that of vascular plants (21).

Most decomposition takes place on the forest floor and in the soil. There is a lot of activity. The mass of fungal and bacterial decomposers per hectare in fertile forest soil has been calculated to be 5 tonnes (21). The parts of fungi that we usually see are those devoted to reproduction and dispersal (Plates 1, 9, 50, 51, 53). These can be mushrooms, such as the poisonously psychotropic fly agaric, *Amanita muscaria* (Plate 53). They can be branching coral fungi (Plate 50). They can be bracket or shelf fungi, which are responsible for most decay of standing dead trees (Plate 1). These reproductive organs release spores to be dispersed by wind, by animals or mechanically. Because their reproductive parts appear only briefly and occasionally, with the rest of their mass being highly cryptic, gaining a complete list of macrofungi for a forest can take years.

MYCORRHIZAL FUNGI

SWAMP GUM OFTEN fails when grown overseas. This is because its seed is not sufficient in itself to ensure the success of a new individual. The spores of a certain type of fungus are also required for successful tree establishment. This type of fungus is known as mycorrhizal, the name deriving from its close association, indeed mutualism, with the root systems of plants. Mycorrhizas increase the supply of nutrients to the plant, being particularly effective with the macronutrients, phosphorus and nitrogen. They also seem to have a protective role, inhibiting disease and toxic chemical damage and increasing drought resistance. The plants supply the fungus with carbohydrates.

TREES, ATMOSPHERE AND CLIMATE

MORE THAN HALF of the land surface of the earth was covered by trees before the development of agriculture just a few thousand years ago. Carbon was not only stored in the woody parts of trees, but also in the forest soil, in the form of organic matter. The permanent destruction of a large proportion of this forest by people has made a major contribution, currently one-quarter, to the increases in the proportion of carbon dioxide in the atmosphere—which, in turn, may make the world a warmer, wetter and less climatically predictable place than it otherwise might be. Forests transpire much more moisture than agricultural land. In some situations, as in the Amazon basin, rainfall has been predicted to decrease as a result of the impact of forest loss on the amount of moisture in the atmosphere.

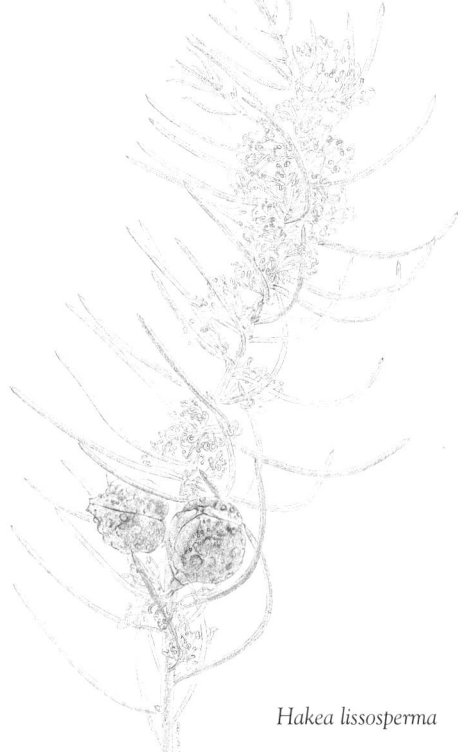

Hakea lissosperma

THE VALUE OF A TREE

IT IS CLEAR THAT the value of a tree is much more than that of the tree itself. Without extensive areas of forest and woodland the world would function very differently and have a much lesser variety of life. This may not worry everyone. There have been many human cultures in which trees have been regarded as enemies of humanity and life has been regarded as in excessive and dangerous variety. There are many human beings alive today who value trees only in their destruction or utilisation, and have no great concern about the current wave of human-induced extinctions. However, the gardens of the world testify to at least equally prevalent positive human feelings about trees. There are few gardens, in western societies at least, that do not possess at least one tree that promises no fruit, no fuel, only itself.

What is it in such practically useless trees that motivates their planting and allows their survival?

FEELINGS ABOUT TREES

MY OWN FEELINGS about trees developed from the contrast between living as a child in a brand new, virtually treeless, suburb of Melbourne, then called Moorabbin, and frequently staying with my grandparents in a well-treed rural village called Macedon. The tree we children loved most at Macedon, among the many we loved, was not planted. It was an ancient manna gum (*Eucalyptus viminalis*) that grew between the dirt road and the cypress hedge that marked the front of my grandparents' property. This spreading tree had one large branch that swept down to almost ground level on which my brother and I would be seated while our uncle moved it as a swing. At dawn magpies would sing their bell flute songs from its upper branches. At unpredictable times of night the screeches of koalas and brush-tailed possums would emerge from its direction. From my grandfather's study desk, with its miniature Egyptian obelisk and First World War shell-casings, we watched the morning train pass behind its foliage. My parents dug up one of its many seedlings to place in the front yard at Moorabbin. It grew quickly, outpacing the liquidamber that was previously the biggest tree. It was climbable during my childhood, unlike any other of our trees. My mother did it once, better than I could, at the time or now. It had roughly fissured bark at its base, chalky white upper branches and sparse, hard green, sickle-shaped leaves through which I could see the grey Melbourne sky.

I loved that tree, eventually punished by death for the crime of dropping large branches in the presence of a fence, but it was a dull love, the sort of muted love I felt for the beach and waters at Black Rock, pale, dirty and artificial compared to the beaches and waters we visited on our summer holidays. The leadenness of living in the tasteless artifice that constituted recently established suburbia could be relieved, but not cured, by a tree. The cure lay for me in nature, rooted deeply in time, in the relative absence of artifice. The time did not need to be overly deep. The harmony, clarity and mystery that allowed me to feel fully alive was experienced at Macedon, in the old pine plantation and the lake full of waterbirds and weeds, with its yellow-green shafts of light and dark green cold depths into which the pine needles fell. The lake and plantation were created by humans, but had been harmonised by natural forces and were rooted firmly in a past that seemed as remote to me, as a child, as the upper branches of the trees that dwarfed my small freckly form.

I feel emotional pain when trees are felled by humans, a pain that increases with the age and size of the tree. I feel no such pain when trees are felled or killed by non-human agency. I know that many others feel the same pain. In the 1960s in Melbourne tree haters were in ascendancy while old age pensioners quietly starved. The now defunct *Herald*, the evening newspaper, ran campaigns on both issues, prompting the University of Melbourne student newspaper, in a satire *Hairoiled* edition, to run the headline: Hungry Pensioners Ringbark Street Trees. Some understanding of the origin of the continuing tension between lovers of trees and those who are dislikers or indifferent may be gained from a brief consideration of the transition of human beings from gatherers and hunters, enmeshed in nature, to rich urbanites, enmeshed in consumer durables.

GATHERERS AND HUNTERS AND TREES

THE WEB OF LIFE is also very much a web of death. Animists see
spirits, akin to those putatively theirs, everywhere in the ecosystem,
spirits perceived by some to be capable of mobility between types of life
and components of the inanimate environment. They see themselves
as part of nature, rather than a species apart. Animist people perceive
themselves as belonging to the land, rather than possessing it.
This does not mean that animists are necessarily ecological geniuses,
are kind to animals or devote their lives to the maintenance of
biological diversity. We will never know, but I suspect that the animist
gatherers and hunters of Australia agonised as little about the
extinction of the Pleistocene carnivorous megafauna as we agonise
over the imminent extinction of smallpox. Kindness to other animals is
not a common attribute of carnivorous species. Prey is killed for the
benefit of the predator, not the victim, much as we rationalise it
otherwise. It is difficult to be empathetic to any great degree if your life
depends directly on killing some other organism.

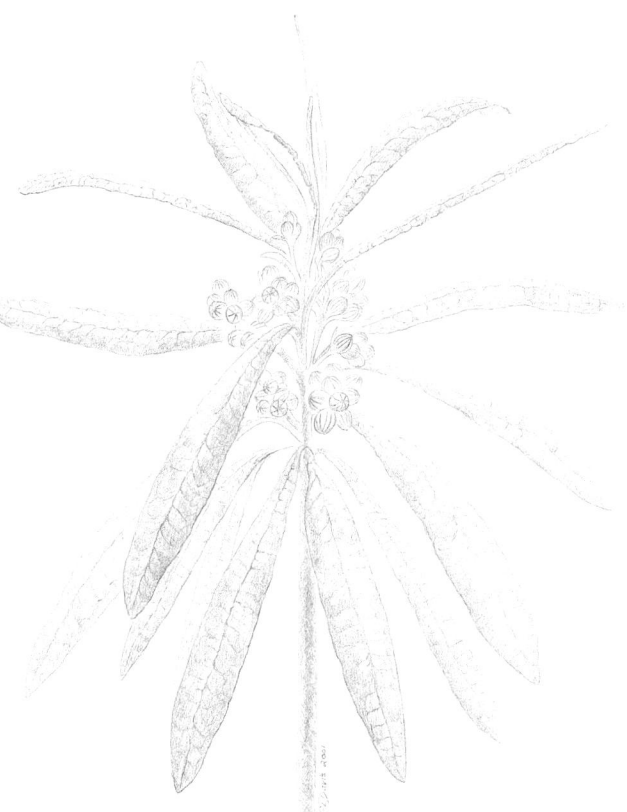

Bedfordia salicina

The current scientifically accepted belief is that people evolved
in Africa and spread as gatherers and hunters to all other continents
except Antarctica. Relative stability of the gatherer and hunter
population and lifestyle seems to have followed an initial period of
invasion of new lands, during which the web of life might have been considerably adjusted, and seems to
have lost quite a few nodes. The waves of animal extinctions that followed the invasion of people into
previously peopleless lands are a matter of some controversy in Australia and North America today,
with some passionately arguing that people were largely or solely responsible for the extinctions, and
others equally passionately arguing that there was no way that hunting and gathering people could have
rendered large animals extinct, and that the extinctions were caused by climatic change (4, 10). The
balance of the argument seems to me to lie on the side of some degree of human cause, given the better
documented extinctions that have followed the invasion of other predators, such as the dingo and fox,
into Australia, and the dramatic changes in vegetation that have occurred where gatherers and hunters
ceased their management of the ecosystem (16).

There is a lot to respect about gathering and hunting cultures, especially if you are fond of trees.
Once the initial shock to the ecosystem of the human invasion was over, gatherers and hunters became
one of the nodes in a remodelled web, in which it was in their interests to maintain a steady flow of
resources. Behaviour patterns that ensure a steady flow of resources are likely to be selected for by the
success of the groups that adopt them, and the failure of those that do not. The behaviours are reinforced
and perpetuated by becoming part of the religious truth and accepted social mores of the group (27). It
has recently been suggested that some of these behaviours may have extended the range of rainforest in

northern Australia, with frequent patch burning at low intensity preventing the intense fires that can eventually destroy this formation (5). Gathering and hunting people lived everywhere, from rainforest to desert. They used the bark, leaves, branches, flowers and fruits of trees, but trees remained standing. Many of the trees that remained standing during the gathering and hunting occupance of Australia are still alive, probably looking little different to their selves in 1788 (e.g. Plates 1, 2, 7, 12–16, 18–22, 24, 25, 29).

The stability and harmony that the romantics of Europe perceived gathering and hunting people to possess was real in a relative sense. Europe was in the early stages of eating up the stored wealth of the world, a process involving rapid growth in human biomass at the expense of other biomass. The gathering and hunting socioeconomy was limited by the potential of people to maintain themselves as one predatory species in an ecosystem, so population levels were relatively constant, as was the impact of our species on the environment. Cultural restrictions on population growth were common in gathering and hunting societies. These seem to have reflected the difficulties of carrying large numbers of young children when moving, and a desire to keep numbers below the carrying capacity of their territory.

THE FALL OF PEOPLE AND TREES

AGRICULTURE DEVELOPED independently in the new and old worlds, and probably independently within both. My favourite theory for the development of agriculture is that it happened by accident, as it is hard to conceive that anyone would voluntarily trade the long, healthy, and not particularly onerous life of gatherers and hunters for the short, insecure and brutish life that has so often been the lot of farmers. A group of gathering and hunting people might have noticed that some of the plant species they consumed established in disturbed ground around their camps. They may have started to give the plants they most preferred a bit of help, by clearing away competitors and deliberately disturbing the ground. They may then have started sowing seed or planting tubers. This activity increased their food supply, enabling a sedentary life as good or better than their previous one. A sedentary life and the extra food allowed a relaxation of cultural restrictions on numbers of children. When the numbers of people grew too large to be supported in the territory where gathering morphed into agriculture, the agricultural people could displace neighbouring less innovative gatherers and hunters by sheer force of numbers. This they did. The more they expanded the more they depended on agriculture for survival, making them much more subject to disease and famine than in their gathering and hunting days.

While some trees became agricultural crops, most were an impediment to agricultural production. Stone axes and fire fought the forest, followed by bronze axes, iron axes, steel axes and bulldozers. In some places the victory of agricultural people over the forest appears to have been brief, with Khmer and Mayan cities being enveloped by reinvading trees. Some people practised shifting cultivation, in which the reinvading forests restored the productive capacity of previously cultivated land, before it was cultivated again. However, in most places the forests were permanently kept at bay from the cleared land. In many agricultural civilisations the remaining forests were generally regarded as threatening places, or were the preserves of the rich, who atavistically hunted game within them.

In 1788 Australia was the last of the inhabited continents to succumb to agriculture, leaving only

Antarctica uncultivated. Many people in the Australian countryside have spent a lot of time keeping trees from invading their productive land, and in destroying trees to create new agricultural and pastoral land. It is not easy stopping trees coming back. Old paintings of Mount Wellington show pastures where now there is dense stringybark forest, and the thriving port town of Pillinger on Macquarie Harbour in western Tasmania is now but crumbling walls from which the myrtle sprout. It is therefore no great mystery why country people are not necessarily overfond of trees that do not contribute to agricultural production.

CIVILISATION, GARDENS AND TREES

CITIES CANNOT ARISE unless the countryside produces more food, fibre and construction material than it uses. Thus, civilisation relies on the appropriation of the agricultural surplus, a fact not necessarily well appreciated in the western cities of today. The use of trees as ornament seems to have arisen among those who most appropriated the surplus, the hanging gardens of Babylon being an early example. Ornamental gardens, and especially those with trees, signify wealth, an ability to put energy into culture rather than just sustenance.

Once, as I flew from Sydney to Melbourne, I looked out of the window and saw an extensive forest on a wide plain running into a large, round lake. I felt disorientated, as I knew that this combination of geographic features did not exist between the two cities. Then, Melbourne and Port Phillip Bay clicked into view. My eyes had been deceived by the low angle of approach and the profusion of trees in suburban gardens. Melbourne is a wealthy new world city in which most people have space to plant trees if they so desire. It is an urban forest, while trees are still scarce in Beijing.

There are many different motivations for planting trees in the urban forest, and its outliers in the gardens of the relatively rich in the countryside. The planting and nurturing of a tree in a garden can be a spiritual act in several modes, including: an act of loving care or 'husbanding' that transcends human mortality; a celebration of the richness of the gifts of a or the god; a symbol of those who sacrificed themselves for their deity or country, as in memorial avenues; a contribution to the healing of the planet. Tree establishment can be a statement of dedication to consumerism and its handmaid, fashion, as in the expensive transplanting of palms. The aesthetic motive is strong among many who plant trees (Plates 46–49, 55). Shelter from wind and summer sun is another common motive (Plate 57). A sense of place is often a strong motive, with people planting trees that reinforce their sense of connection with their beloved local landscape. Certain trees have emotional associations, reminding those who plant them of places they lived in and loved, as with the oaks and elms that surround the colonial mansions in the countryside of Tasmania (Plate 52). Others make an ideological statement, as in the permacultural garden. A few people collect them. Many people just love trees.

Why do trees dominate suburbia in one part of big new world cities and not others? The answer is somehow related to socioeconomic status. The richer the suburb, the greater the density of trees. This cannot be related to the expense of establishing trees. Even if it were not negligible, and it is much less than that of the concrete borders that are characteristic of treeless gardens, the problem in any area that is environmentally suited to tree growth is keeping out trees, not establishing them. I frequently weed out seedlings of both local native trees and exotic trees, the seeds of which have been generously distributed to my garden by both native and exotic birds. Many of the existing trees in my garden have not required a visit to a nursery. A neglected garden will quickly turn into a thicket of trees. People with treeless gardens are actively keeping trees out.

I suspect that the answer may at least partly lie in social history and psychological factors. The poorer suburbs of western cities have a much higher proportion of people recently displaced from rural areas than do the richer suburbs. The experience of people who have lived their lives in cities is quite different from that of country people. Long term urbanites see trees as very much at the mercy of people, which they obviously are where there are many people, relatively few trees and an active real estate industry. The good side of trees is much more apparent than the bad.

The well-off have more options than the relatively poor. They are generally more in control of their lives, having not only greater income, but also greater autonomy and responsibility in their work. They may therefore not feel as great a need to control nature as those less fortunate. Trees are much less controllable than lawn, concrete, annuals and standard roses. Their roots get under paths and into drains, their fallen leaves make an awful mess, either constantly or once a year clogging up gutters and killing lawn. They attract messy birds. Best not to plant them, or only have those of predictable shape and smaller size that do not drop a lot of messy leaves. Small conifers of many a hue are popular for these reasons.

A garden is not a garden if it is full of plants that have grown without any human interference, and equally not a garden if bereft of plants altogether. It is difficult, if not impossible, to have a garden totally under human control. While some heavily controlled gardens have an architectural attractiveness, the gardens that I find most pleasing are those in which nature plays a great part, or is simulated to play a great part. However, I am probably not a great judge of gardens, in that I have yet to see one—even my own, which I love—that approaches in beauty and harmony the forests and woodlands of nature.

Jamie Kirkpatrick is Professor of Geography and Environmental Studies
at the University of Tasmania.

Athrotaxis selaginoides

BIBLIOGRAPHY

This bibliography includes both the sources used in writing the book, and other publications that might interest those interested in trees and the ecosystems and gardens in which they occur. Literature relevant to the forests and woodlands depicted in this volume is given particular emphasis.

(1) Adam, P., 1992. *Australian Rainforests.* Clarendon Press, Oxford.

(2) Attiwill, P. M., and Leeper, G. W., 1987. *Forest Soils and Nutrient Cycles.* Melbourne University Press, Melbourne.

(3) Barker, P., and Kirkpatrick, J. B., 1994. *Phyllocladus aspleniifolius:* variability in the population structure, the regeneration niche and dispersion patterns in Tasmanian forests. *Aust. J. Bot.* 42, 163–190.

(4) Bowman, D. M. J. S., 1998. Tansley review 101: the impact of Aboriginal landscape burning on the Australian biota. *New Phytologist* 140, 385–410.

(5) Bowman, D. M. J. S., 2000. *Australian Rainforests—Islands of Green in a Sea of Fire.* Cambridge University Press, Cambridge.

(6) Cullen, P. C., and Kirkpatrick, J. B., 1988. The ecology of *Athrotaxis* D. Don (Taxodiaceae) I. Stand structure and regeneration of *A. cupressoides. Aust. J. Bot.* 36, 547–60.

(7) Dombrovskis, P., Brown, B., and Kirkpatrick, J. B., 1998. *Dombrovskis.* West Wind Press, Hobart.

(8) Dombrovskis, P., Flanagan, R., and Kirkpatrick, J. B., 1996. *On the Mountain.* West Wind Press, Hobart.

(9) Enright, N. J., and Hill, R. S., 1995. *Ecology of the Southern Conifers.* University of Melbourne Press, Melbourne.

(10) Flannery, T., 1994. *The Future Eaters.* Reed Books, Melbourne.

(11) Gilfedder, L., 1988. Factors influencing the maintenance of an inverted *Eucalyptus coccifera* treeline on the Mt. Wellington Plateau, Tasmania. *Aust. J. Ecol.* 13, 495–503.

(12) Hassall, D. C., and Kirkpatrick, J. B., 1985. The diagnostic value and host relationships of the dependent synusia in the forests of Mount Korobaba, Fiji. *N.Z. J. Bot.* 23, 47–54.

(13) Hobbs, R. J., and Yates, C. J. (eds.), 2000. *Temperate Eucalypt Woodlands in Australia: Biology, Conservation, Management and Restoration.* Surrey Beatty and Sons, Chipping Norton.

(14) Jarman, S. J., and Kantvilas, G., 1995. Epiphytes on an old Huon Pine tree (*Lagarostrobus franklinii*) in Tasmanian rainforest. *N.Z. J. Bot.* 33, 65–78.

(15) Kirkpatrick, J. B. (ed.), 1991. *Tasmania's Native Bush: A Management Handbook* Tasmanian Environment Centre, Hobart.

(16) Kirkpatrick, J. B., 1999. *A Continent Transformed—Human Impact on the Natural Vegetation of Australia.* 2nd ed. Oxford University Press, Sydney.

(17) Kirkpatrick, J. B., and Hassall, D. C., 1985. The vegetation and flora along an altitudinal transect through tropical forest at Mount Korobaba, Fiji. *N.Z. J. Bot.* 23, 33–46.

(18) Low, T., 1999. *Feral Future—the Untold Story of Australia's Exotic Invaders.* Viking, Ringwood.

(19) McKenny, H., and Kirkpatrick, J. B., 1999. The role of fallen logs in the regeneration of tree species in Tasmanian mixed forest. *Aust. J. Bot.* 47, 745–753.

(20) Pollan, M., 1991. *Second Nature—a Gardener's Education.* Dell Publishing, New York.

(21) Raven, P. H., Evert, R. F., and Eichhorn, S. E., 1986. *Biology of Plants.* Worth, New York.

(22) Reid, J. B., Hill, R. S., Brown, M. J., and Hovenden, M. J., 1999. *Vegetation of Tasmania.* Australian Biological Resources Survey, Canberra.

(23) Shapcott, A., Brown, M. J., Kirkpatrick, J. B., and Reid, J. B., 1995. Stand structure, reproductive activity and sex expression in Huon pine (*Lagarostrobus franklinii* Hook f. Quinn). *J. Biogeogr.* 22, 1035–1045.

(24) Timms, P., 2001. *Making Nature—Six Walks in the Bush.* Allen and Unwin, Crows Nest.

(25) Whitmore, T. C., 1998. *An Introduction to Tropical Rainforest.* 2nd ed., Oxford University Press, Oxford.

(26) Williams, J. E., and Woinarski, J. C. Z. (eds.), 1997. *Eucalypt Ecology: Individuals to Ecosystems.* Cambridge University Press, Cambridge.

(27) Yibarbuk, D., et al., 2001. Fire ecology and Aboriginal land management in central Arnhem Land, northern Australia: a tradition of ecosystem management. *J. Biogeogr.* 28, 325–343.

Renaissance

Once again I am aware
of the benevolence of elms,
that silent loosening of bonds
until the intermittent fall
of autumn's text,
letter by letter,
landing unread on winter coats,
or soaring high upon the wind
as if, for one brief moment,
summer might reinvent itself,
or washed down gratings
to a sea which never shares
its brown and yellow secrets.

Only young children
rustle through the leaves
and then smile up, open-mouthed,
as if they would swallow it all,
speak with the tongues of trees.

Christina Kirkpatrick

PETER DOMBROVSKIS

TEMPERATE FOREST

1. KING BILLY PINE FOREST, SOUTHWEST TASMANIA

2. KING BILLY PINE, SOUTHERN RANGES, TASMANIA

3. MOSS-COVERED TRUNKS, CRADLE MOUNTAIN–LAKE ST CLAIR NATIONAL PARK, TASMANIA

4. CLOUD FOREST, MOUNT ANNE, SOUTHWEST TASMANIA

5. KING BILLY PINE AND PANDANI FOREST, MOUNT ANNE, TASMANIA

6. MOSS AND LICHEN IN MYRTLE FOREST, CRADLE MOUNTAIN–LAKE ST CLAIR NATIONAL PARK, TASMANIA

7. OPEN MYRTLE FOREST, CRADLE MOUNTAIN–LAKE ST CLAIR NATIONAL PARK, TASMANIA

8. MORNING MIST IN MYRTLE FOREST, CRADLE MOUNTAIN–LAKE ST CLAIR NATIONAL PARK, TASMANIA

10. SASSAFRAS TREES IN MIST, MOUNT ANNE, SOUTHWEST TASMANIA

11. TWIN TRUNKS, DECIDUOUS BEECH, CRADLE MOUNTAIN–LAKE ST CLAIR NATIONAL PARK, TASMANIA

12. DECIDUOUS BEECH FOREST, CRADLE MOUNTAIN–LAKE ST CLAIR NATIONAL PARK, TASMANIA

13. DECIDUOUS BEECH ON CONGLOMERATE, CRADLE MOUNTAIN–LAKE ST CLAIR NATIONAL PARK, TASMANIA

14. DECIDUOUS BEECH IN WINTER, CRADLE MOUNTAIN–LAKE ST CLAIR NATIONAL PARK, TASMANIA

15. WEATHER-BLEACHED DECIDUOUS BEECH, CRADLE MOUNTAIN–LAKE ST CLAIR NATIONAL PARK, TASMANIA

16. DECIDUOUS BEECH NEAR BARN BLUFF, CRADLE MOUNTAIN–LAKE ST CLAIR NATIONAL PARK, TASMANIA

17. AUTUMN SNOW, CRADLE MOUNTAIN–LAKE ST CLAIR NATIONAL PARK, TASMANIA

18. MYRTLE TREE, CRADLE MOUNTAIN–LAKE ST CLAIR NATIONAL PARK, TASMANIA

19. SNOW ON PENCIL PINE, CRADLE MOUNTAIN–
LAKE ST CLAIR NATIONAL PARK, TASMANIA

20. SNOW-LADEN SNOWGUM, WELLINGTON RANGE, TASMANIA

21. FROZEN WATERFALL, WALLS OF JERUSALEM NATIONAL PARK, TASMANIA

24. MYRTLES NEAR STREAM, CENTRAL HIGHLANDS, TASMANIA

25. SNOWGUM, CRADLE MOUNTAIN–LAKE ST CLAIR NATIONAL PARK, TASMANIA

26. BARK DETAIL OF SNOWGUM, CRADLE MOUNTAIN–LAKE ST CLAIR NATIONAL PARK, TASMANIA

27. BANKSIA SEED PODS, TASMANIA

28. CABBAGE GUM BOLES NEAR LAKE ST CLAIR, TASMANIA

TROPICAL FOREST

32. COCONUT PALM, HINCHINBROOK ISLAND, QUEENSLAND

33. FOREST VINE, FIJI

34. PANDANUS, FRASER ISLAND, QUEENSLAND

35. EDGE OF MANGROVE SWAMP,
FRASER ISLAND, QUEENSLAND

36. TROPICAL FOREST, DAINTREE, QUEENSLAND

37. FOREST BUTTRESS, FIJI

38. TROPICAL FOREST, FIJI

39. TROPICAL RAINFOREST, SARAWAK, BORNEO

40. TROPICAL VINE, FIJI

41. BANYAN TREE, FIJI

42. MANGROVES AT SUNSET, FRASER ISLAND, QUEENSLAND

URBAN FOREST

44. BAMBOO FOREST, BOTANIC GARDENS, MELBOURNE, VICTORIA

46. *PIERIS JAPONICA* AND SILVER BIRCH, FERN TREE, TASMANIA

47. AUTUMN, LALLA, TASMANIA

48. MAPLES, LALLA, TASMANIA

49. MAPLES, LALLA, TASMANIA

50. CORAL FUNGUS (RIMARIA SP), SOUTHWEST TASMANIA

53. AMANITA MUSCARIA, TASMANIA

54. EUCALYPT IN SNOW, FERN TREE, TASMANIA

55. *RHODODENDRON FALCONERI AND EUCALYPTUS REGNANS, FERN TREE, TASMANIA*

56. GUNNERA MANICATA AND EUCALYPT TRUNKS, FERN TREE, TASMANIA

57. MONTEREY PINE, DERWENT VALLEY, TASMANIA